It is not in the stars to hold our destiny but in ourselves.
William Shakespeare

To Ombline, Côme, Fabiola
I can't wait to live the future you will shape for us!

DIGITAL ATELIER MANIFESTO

By 2025, a $1000 laptop would have the computing power of a human brain and in the near future 40% of US jobs could be automated.

Do you feel the same sense of urgency to get your digital natives ready for a digitally transformed future?

Here is the Digital Atelier manifesto:

Our mission is to awaken the potential of Digital Natives.

We acknowledge the acceleration of technology that will trigger a singularity redefining our economy, social structure, human relationships, jobs, education, and entertainment.

We want our kids to blossom, be happy. It takes more than just coding skills. Self-awareness, critical thinking, human values, being in touch with reality will be paramount in a world where artificial intelligence and robots outperform humans.

We want kids to become shapers. They leant back watching TV, then leant forward engaging with mobile applications and social networks. We would like them to make, create, think, feel.

We strive to develop an experience that stimulates the natural curiosity of Digital Natives. This is not about linear learning. Don't expect quiz to validate your learning. We guide you through the maze of content to build diverse skills and encourage you to dive in topics you show interest for.

We want to trigger discussions between digital natives (teenagers) and digital immigrants (parents). Digital Atelier is about strengthening human relationships. Our success is that you get to see the sparkle in each other eyes.

We curate existing content. Everything is out there. What matters is to structure content, combine it, and find the entry point that would trigger Digital Natives' interest.

We developed a 4 skill model:

- Make: we believe coding is a language, we want you to become comfortable searching within a code, identify where you could hack a code. Coding goes hand in hand with robotics. 3d print and other making magic complement this dimension.

- Create: this is about designing, human centric innovation. Algorithm is a big part of making things happen in a digital world. Go beyond Algebra – welcome to Algorithmic.

- Think: develop critical thinking, raise your self-awareness, think through the social impact of the digital transformation.

- Feel: technology made us more efficient, more productive. As technology will become more intelligent, it will now impact relationships, feelings. Let's celebrate emotions, they make us humans!
Do you want more? Join us on our journey, leave us an email on digitalatelier.org.

We are so excited to share what we found … Give us feedback! benoitgaillard@digitalatelier.org

Enjoy!
Benoît and Yves

Download the **Unitag QR code scanner** app to access the videos and articles featured in this book.

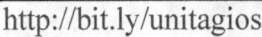

http://bit.ly/unitagios

http://bit.ly/unitagandroid

Meet the editorial board

Benoît Gaillard

Digital Transformation
Manager at Caterpillar

Founder
digitalatelier.org

Yves Zieba

Strategy and Innovation
Advisor
Based in Geneva

Project Director
digitalatelier.org

"Pranav Mistry". Source wikipedia – Public domain

Make

Break through the 5x10 screen. Don't live through a screen, but rather make the real world your interface. **To access content, flash this image with the Unitag QR code scanner app.**

"Spare_Parts_poster" by source. Wikipedia – Fair Use poster

Make

How 4 Mexican immigrant kids and their cheap robots beat the MIT. This movie is a great discussion starter to get your digital natives into making.

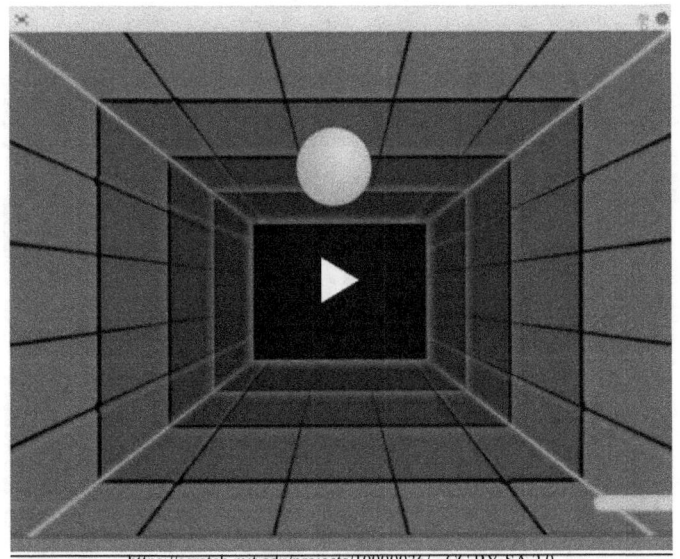

Make

Hack the pong game! Make it multiplayer.
No need to learn how to code. We show you where and
how to amend the code in this 2 minute video. Easy!

Create

When robotics meet Art ... a new type of expression sees the light. Check out light painting by Thymio!

"Ex-machina-uk-poster" by Source (WP:NFCC#4). Wikipedia – Fair Use poster.

Think

When does a computer become an Artificial Intelligence?
Watch the movie !

Dean Kamen by Jason Gessner – flickr CC BY-SA 2.0

Create

Meet da Vinci reborn… on a mission to deliver drinkable water to the world with his newest invention Slingshot.

"Farmer, Nicaragua" by Neil Palmer (CIAT) – CC BY-SA 2.0 flickr.

Think

Data based decision making is a critical skill.
Test your analytical thinking skills! Will you pass the test?

« Her2013Poster » by Wikipedia – fair use

Feel

A man falls in love with an Artificial Intelligence. Is this already reality?

Feel

Spelling is hot on dating websites! Actually, women
judge correct spelling second in importance after
personal hygiene.

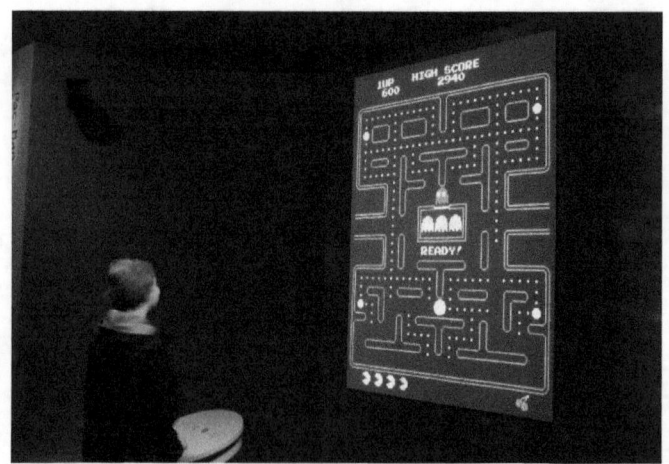

Make

Hack a gaming monument! Bring in some friends to go after those ghosts!

Think

Statistical Prejudice! 'Inequity gets baked in'. Class and race struggles get entwined with technology.

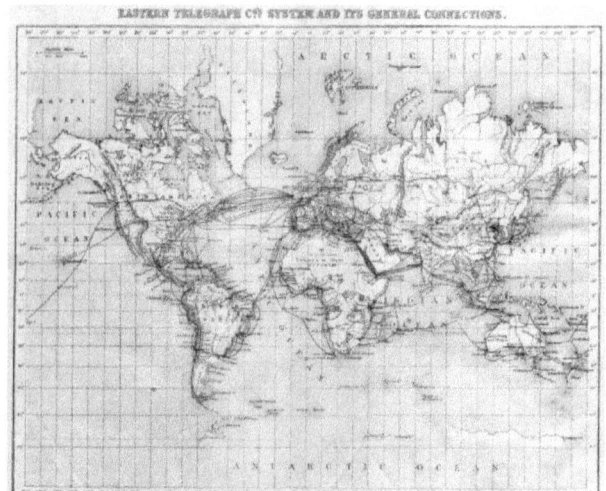

A.B.C. Telegraphic Code 5th Edition, via Atlantic-cable by Wikipedia. Public domain.

Make

30% of kids confuse a browser (Internet Explorer, Chrome, Safari) and the internet. Discover the physical reality behind the 'web'.

Feel

Pi … a math mystery becomes a melody! Listen.

"Back to the Future" by Source. Licensed under Fair use via Wikipedia
"Internet Archive Wayback Machine logo" by Internet Archive (IA) -
archive.org/web/

Think

Try the Internet Wayback Machine and see how your
preferred website was looking like 10 years ago.

Make

Learn from Mother Nature ... meet stickybot and gecko inspired robots. Discover some surface magical forces.

"Shattering silence" by Phil Roeder. Art by James Ellwanger. Flickr. CC BY 2.0

Feel

Would an Artificial Intelligence come up with the aesthetical intelligence of an oxymoron? Why would you craft a figure of speech juxtaposing apparently contradictory elements to reveal a paradox?

Cubelets by Carl – Flickr. CC BY-ND 2.0

Make

The challenge: design a robot that would adapt its shape to the task or the surface in which it evolves. Discover why it matters.

Eva Official US Release Trailer 1 (2015) by Movieclips Film Festivals & Indie Films. Youtube.

Feel

What would be the one question that would ultimately differentiate between humans and robots? "What do you see when you close your eyes?"

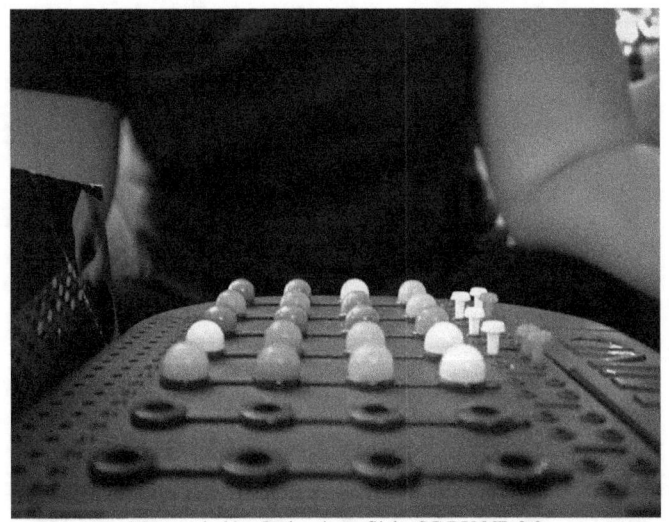

Mastermind by Owlpacino - flickr CC BY-ND 2.0

Create

Mastermind really? Challenge yourself!

Techcrunch via flickr CC BY 2.0

Create and Think

Solving algebra? There is an app for that! Learn about algorithm and why it matters.

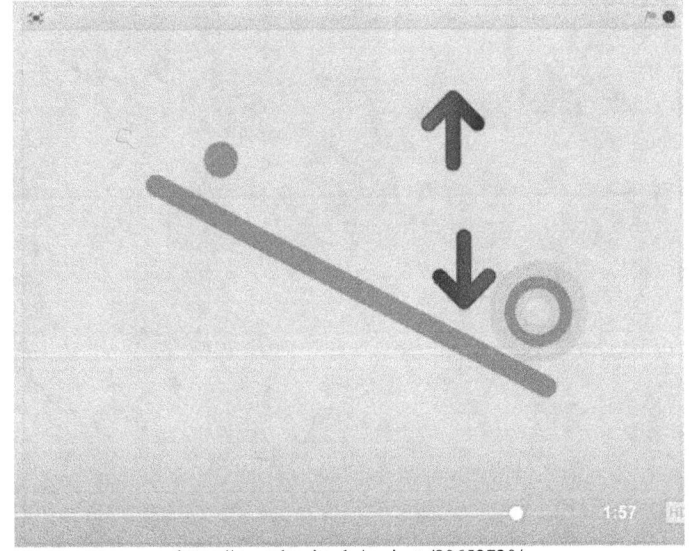
https://scratch.mit.edu/projects/80652720/

Make

Add a twist to Tilted, bring in enemies preventing you
from reaching your goal.

Create

Learn how to run your own business. Invest, Price, Forecast, Capitalize … A 'refreshing' game to experience the basics of entrepreneurship.

Make and Create

Welcome to the new farms cityFARM! Could you digitalize food and fix the global nutrition problem? The MIT is working on it… most important; agriculture is cool again now that it is managed from your iphone!

https://scratch.mit.edu/projects/80652720/

Make

Create your own new levels in this game. Easy!

Create

Copying nature! Check out how Harvard brains combined smart robotics and algorithm allowing robots to swarm.

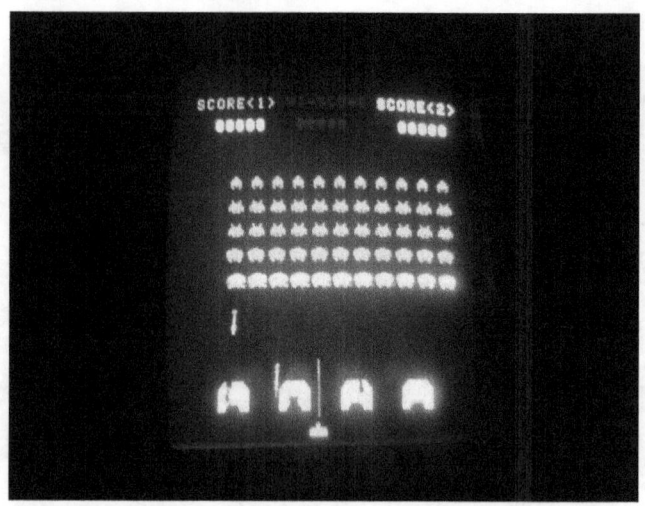

Make

Shoot them up! Rediscover a classic video game. How does the program actually know when the bullet hits the enemy?

Make

Thymio rocks to learn robotics! Proudly developed in
Switzerland at EPFL. Unleash your creativity: Thymio.org

« Watson_Jeopardy ». Charlie Wollborg via Flickr CC BY-SA 2.0

Create

Watson – the artificial intelligence by IBM wins
$1'000'000 in Jeopardy challenge. Easy? Well, 6 years
of research … here summarized in 20 minutes.

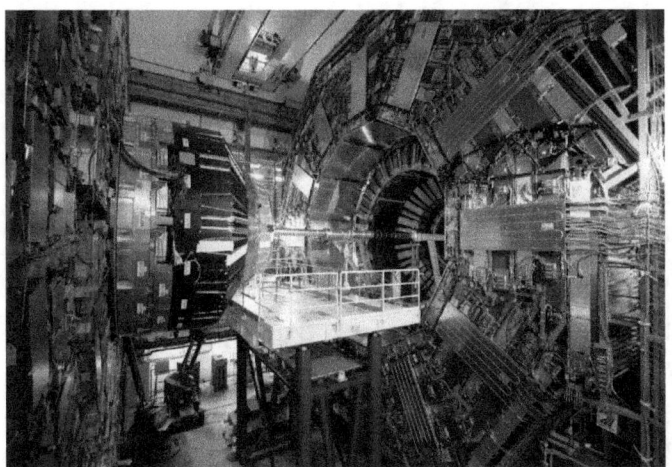
"Particle Fever" by http://www.imdb.com/title/tt1385956/. Via Wikipedia

Think

More dramatic than a thriller, better gadgets than in James Bond's movies … Discover the largest scientific experimentation of all times. Experience the thrill of early failures and success! CERN discovers God's particle!

Feel

Is Twitter the new Haiku?
Haiku 17 syllables fit well the 140 characters of twitter.
Discover how Pr. H.W. Banks tells US history tweeting
haiku.

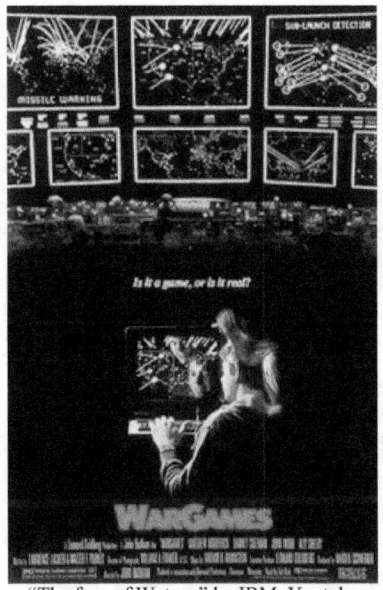
"The face of Watson" by IBM. Youtube

Create

Is it a game or is it real?
Go back in times and experience dial-up modem, floppy
disks. A great tale of hacking and Artificial Intelligence.

Make

Hack this shoot them up game, create a secret weapon.
Destroy the enemy by the sheer power of your voice!

Thymio meets ECAL from Vimeo by Arthur Touchais. ECAL

Make and Feel !

When technology meets design ... you get magic!

avana public artwork – empty headed conversation
by Gareth Williams on Flick CC BY 2.0

Feel

Is text messaging killing the 'Art of conversation'?

« Coding Gouter » By TEDxYouth. Youtube

Coding

Pour nos amis francophones … le coding goûter, la programmation racontée de manière convaincante à des enfants de 10 ans!

Think

Learn how to become critical about Google search results. 10 points to check if a website is credible!

Create

Video streaming with too many geeks sucking my home wifi bandwidth leads to pixelated and ghostly graphics. Why?

Think

Watson - the computer that won Jeopardy in 2011 was shrinked in 2014 to the size of a shoebox.

United States Federal government, released into the public domain - wikipedia

Create

CIA agents need you. Will you break the code and save the President?

Make

Operate heavy equipment … from your bedroom. Ans shows you in this 2 minute video how to twist the game to make it more dramatic and fun?

"TheMusicOfThePrimes" by Source. Licensed under Fair use via Wikipedia

Create

Prime numbers hold tight to their mystery. Get inspired and start competing taking your school to the top of the leaderboard.

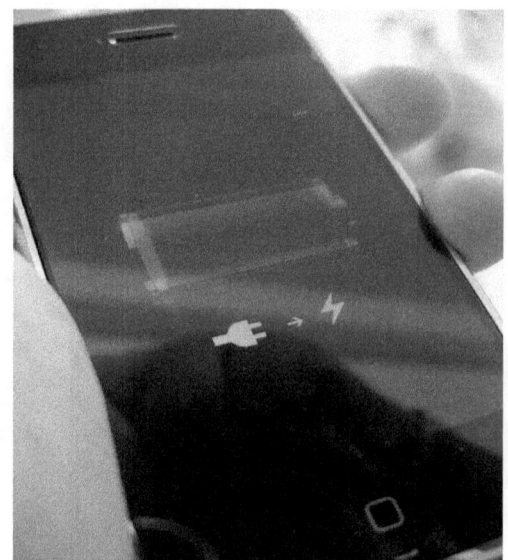

Think

How much energy does your iPhone actually consume?
As our always-on culture keeps growing, we need to
become aware that digital does come with a utility bill!

Create

Minecraft uses an algorithm to generate the virtual world as you progress through it. This algorithm has limitations. It becomes creative beyond what the developer could imagine when you reach that far … welcome to Far Lands!

Maurizio Pesche via Flickr CC BY-2.0

Create

Smart and cheap innovation. Virtual reality democratized!
Check out how this is used to transform tourism, extreme sport, real estate, tradeshow …

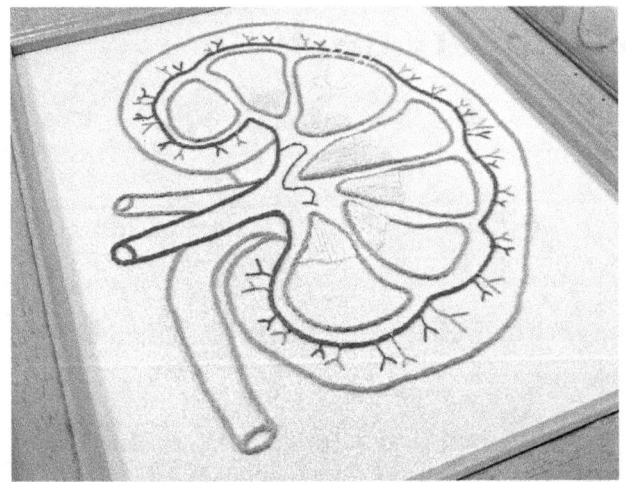

Create and Think

A solution to the shortage of organ donors? 3D printing organs is now a reality! 'Bio-designers' started to imagine new features for our body to augment humans … Time to think of bio-ethics!

Some more inspiration ...

TED.org – ideas worth sharing delivered in 10' conferences

scratch.mit.edu – coding made simple and entertaining. No bug. Your kids would love it.

camp.withgoogle.com – fun science activities led by experts. Every day is a camp day! Google at its best.

Khan academy – that largest school with 10 million students learning from short videos!

diy.org – everyday is a camp day! Great video tutorials.

thymio.org – an active and creative community. Daily inspiration for robotics. Proudly developed at EPFL.

What's next?

This book is a first step in our journey to awaken Digital Natives to their full potential. We have a plan to develop a mobile application that allows you consuming the Digital Atelier content through one unified interface. Like – Dislike – Share – Archive our content. Digital Atelier's algorithm selects which content is most likely to interest the kids based on history, context, content sequence.

Help us identify what features would be most helpful for you. Vote here by flashing the image on the previous page with the unitag app.

Every book purchased allows curating one more piece of content published on the Digital Atelier.

Thanks!

Thanks to you. Offer the book, help spread knowledge!

Thanks to those contributing to the Digital Atelier:

Rilana – our amazing remote assistant
http://www.maika-assistance.com/

Ans – the scratch code hacker. The world is flat!
https://www.elance.com/s/visartstudios/

Alexis from **Unitag.io** – augmenting our book with
entrepreneurial spirit and reliable technology.

Our iconic sparkling girl in the cover page by
http://www.morgansessions.com/s